BEASTLY BATTLES

SIBERIAN TIGER vs SABRE-TOOTHED CAT

BY CHARLES C. HOFER

a Capstone company — publishers for children

Raintree is an imprint of Capstone Global Library Limited, a company incorporated in England and Wales having its registered office at 264 Banbury Road, Oxford, OX2 7DY – Registered company number: 6695582

www.raintree.co.uk
myorders@raintree.co.uk

Hardback edition text © Capstone Global Library Limited 2025
The moral rights of the proprietor have been asserted.

All rights reserved. No part of this publication may be reproduced in any form or by any means (including photocopying or storing it in any medium by electronic means and whether or not transiently or incidentally to some other use of this publication) without the written permission of the copyright owner, except in accordance with the provisions of the Copyright, Designs and Patents Act 1988 or under the terms of a licence issued by the Copyright Licensing Agency, 5th Floor, Shackleton House, 4 Battle Bridge Lane, London, SE1 2HX (www.cla.co.uk). Applications for the copyright owner's written permission should be addressed to the publisher.

Editorial credits
Edited by: Aaron Sautter
Designed by: Bobbie Nuytten
Media Research by: Rebekah Hubstenberger
Production by: Whitney Schaefer
Originated by Capstone Global Library Ltd

ISBN 978 1 3982 5607 1 (hardback)

British Library Cataloguing in Publication Data
A full catalogue record for this book is available from the British Library.

Acknowledgements
We would like to thank the following for permission to reproduce photographs: Alamy: Benny Marty, 5 (bottom), Stocktrek Images, Inc., 17, 19, 27 (top), 23, 24, Universal Images Group North America LLC/DeAgostini, 11; Getty Images: Ibrahim Suha Derbent, 5 (top), iStock/Andyworks, Cover (top), 13, iStock/Freder, 20, iStock/Kantapatp, 28, iStock/ViktorCap, 8, Kathleen Reeder Wildlife Photography, 21, Mike Hill, 15; Shutterstock: Daniel Eskridge, Cover (bottom), GUDKOV ANDREY, 7, Kutikova Ekaterina, 9, Nicolas Primola, 6, Sergey Uryadnikov, 27 (bottom), Stanislav Duben, 25, Vac1, 29

Every effort has been made to contact copyright holders of material reproduced in this book. Any omissions will be rectified in subsequent printings if notice is given to the publisher.

All the internet addresses (URLs) given in this book were valid at the time of going to press. However, due to the dynamic nature of the internet, some addresses may have changed, or sites may have changed or ceased to exist since publication. While the author and publisher regret any inconvenience this may cause readers, no responsibility for any such changes can be accepted by either the author or the publisher.

Printed and bound in India.

CONTENTS

READY ... SET ... FIGHT! 4
BEASTLY BRAWLERS 6
ONE BIG KITTY. 8
TERRIBLE TEETH .10
A GIANT BITE 12
KILLER CLAWS. 14
NO ESCAPE 16
SUPER-SIZED SABRE-TOOTH 18
A POWERFUL PREDATOR 20
SNEAK ATTACK. 22
SLY HUNTERS 24
FELINE FIGHT! 26
WHO'S THE WINNER? 28

GLOSSARY. .30
FIND OUT MORE. 31
INDEX 32

Words in **bold** are in the glossary.

READY... SET... FIGHT!

Are you ready for a battle between big cats? These ferocious **felines** are ready to fight.

Sabre-toothed cats lived millions of years ago. They were the top cats of their time.

Siberian tigers are big and powerful too. Which fierce beast will win?

BEASTLY BRAWLERS

The Siberian tiger and sabre-toothed cat have a lot in common. They're both big, bad cats. They have sharp claws and big teeth. They're expert **predators**. They use their speed and strength to catch **prey**.

ONE BIG KITTY

Siberian tigers are huge. They're the world's largest cat. They're bigger than jaguars or panthers. They're even bigger than African lions.

Most pet cats weigh about 4.5 kilograms.

But a Siberian tiger can weigh up to 408 kg.

That's one *big* kitty!

TERRIBLE TEETH

The sabre-toothed cat is known for its huge teeth. Its **canine teeth** grew up to 20 centimetres long. They were curved like a sword called a sabre. The cat's long teeth might have helped to cut **flesh** from its prey.

A GIANT BITE

The Siberian tiger has large teeth too. Its canines can be up to 7.2 centimetres long.

These tigers are known for their powerful bite. Their jaws are so strong that they can snap bones.

The sabre-toothed cat had a much smaller jaw. The cat's bite was dangerous. But it was weak compared to the Siberian tiger.

KILLER CLAWS

The Siberian tiger has sharp claws — and they're huge! They can be up to 10 centimetres long. Their claws help to capture and hold prey.

Like all cats, the tiger's claws are **retractable**. This helps keep them sharp and deadly.

NO ESCAPE

The sabre-toothed cat also had big, sharp claws. They were about the same size as the Siberian tiger's claws. The sabre-toothed cat used its claws to catch prey like bison and camels.

The sabre-toothed cat had strong front legs. They helped it hold on to its prey. There was no escaping a sabre-tooth's powerful claws and legs.

SUPER-SIZED SABRE-TOOTH

Sabre-toothed cats were huge. It was one of the largest hunters of its time. The sabre-toothed cat weighed up to 340 kilograms.

Sabre-toothed cats had to be big. They competed for food with other big hunters. These included the dire wolf and short-faced bear.

A SABRE-TOOTH CAT DEFENDS ITS PREY FROM DIRE WOLVES.

A POWERFUL PREDATOR

The Siberian tiger likes to **ambush** its prey. It hides in the forest. When the prey walks by – it pounces!

Tigers can run up to 97 kilometres per hour. This helps them to chase and catch their prey. Not many animals can escape the tiger's attack.

SNEAK ATTACK

Sabre-toothed cats probably **stalked** their prey. They followed prey slowly. The big cats stayed quiet and hid. Then, at the right moment, they attacked! With their powerful claws and large teeth, other animals didn't stand a chance.

SLY HUNTERS

Sabre-toothed cats were **social** animals. They may have hunted in packs, like wolves do today. They could have worked together to kill large prey. This **strategy** probably helped the big cats survive.

Siberian tigers are **solitary** cats. They hunt alone. They don't like being around other tigers.

FELINE FIGHT!

Five fierce fighters face off deep in a forest. A Siberian tiger is surrounded by several hungry sabre-toothed cats. The sabre-tooths growl and hiss. They show off their long teeth.

But then the tiger responds. **ROAR!** This tells its enemies, "Back off!"

It's time for a big cat battle!

WHO'S THE WINNER?

You've met two big killer cats. The Siberian tiger is a powerful predator. The sabre-toothed cat is an ancient master hunter. Which beast would win this clash of big cats?

	Siberian tiger	**Sabre-toothed cat**
HABITAT	mountain forests of Asia	forests and grasslands of North America
WEIGHT	408 kg	340 kg
LENGTH	up to 3.7 m	about 1.7 m
SPEED	97 kph	unknown
WEAPONS	• teeth, 7.6 cm long • claws, 10 cm long • very strong bite	• teeth, 20 cm long • claws, 10 cm long • strong legs
DEFENCES	sharp claws, speedy runner	sharp claws, lived in groups
STRATEGY	stalks prey alone	hunted in packs

GLOSSARY

ambush surprise attack

canine teeth pointed teeth that help tear food into smaller pieces

feline any animal of the cat family

flesh soft parts of the body that cover the bones, such as skin and muscles

predator animal that hunts other animals for food

prey animal that is hunted and eaten by another animal

retractable able to slide in and out

social living in groups or packs

solitary living and hunting alone

stalk hunt an animal in a quiet, secretive way

strategy plan or method of achieving a goal

FIND OUT MORE

BOOKS
Bengal Tigers (Animals in Danger), Nancy Dickmann (Black Rabbit Books, 2019)

Extraordinary Dinosaurs and Other Prehistoric Life Visual Encyclopedia, DK (DK Children, 2022)

Tigers (Animals), Jaclyn Jaycox (Raintree, 2023)

WEBSITES
kids.kiddle.co/saber–toothed_cat
Read more about the sabre-toothed cat.

www.bbc.co.uk/newsround/22831701
Watch rare footage of Siberian tigers in the wild.

www.wwf.org.uk/learn/fascinating-facts/tigers
Discover more tiger facts on the World Wildlife Fund website.

INDEX

dire wolves 18

pet cats 9

sabre-toothed cats
  bite 12
  claws 6, 16, 22, 29
  habitat 29
  hunting strategies 22, 24, 29
  legs 16, 29
  prey 6, 10, 16, 19, 22, 24
  range 29
  size 18, 29
  speed 6, 29
  teeth 6, 10, 22, 26, 29
  time on Earth 4

short-faced bears 18
Siberian tigers
  bite 12, 29
  claws 6, 14, 29
  habitat 29
  hunting strategies 20–21, 25, 29
  prey 6, 14, 20, 21, 29
  range 29
  size 8–9, 29
  speed 6, 21, 29
  teeth 6, 12, 29

ABOUT THE AUTHOR

Charles C. Hofer is a biologist and writer living in New Mexico. He loves cats. However, he's glad he doesn't live with the deadly beasts featured in this book!